I0469304

# Interstate Bank Building Fire
# Los Angeles, California

Investigated by: J. Gordon Routley

This is Report 022 of the Major Fires Investigation Project conducted by TriData Corporation under contract EMW-8-4321 to the United States Fire Administration, Federal Emergency Management Agency.

Department of Homeland Security
United States Fire Administration
National Fire Data Center

# U.S. Fire Administration Fire Investigations Program

The U.S. Fire Administration develops reports on selected major fires throughout the country. The fires usually involve multiple deaths or a large loss of property. But the primary criterion for deciding to do a report is whether it will result in significant "lessons learned." In some cases these lessons bring to light new knowledge about fire--the effect of building construction or contents, human behavior in fire, etc. In other cases, the lessons are not new but are serious enough to highlight once again, with yet another fire tragedy report. In some cases, special reports are developed to discuss events, drills, or new technologies which are of interest to the fire service.

The reports are sent to fire magazines and are distributed at National and Regional fire meetings. The International Association of Fire Chiefs assists the USFA in disseminating the findings throughout the fire service. On a continuing basis the reports are available on request from the USFA; announcements of their availability are published widely in fire journals and newsletters.

This body of work provides detailed information on the nature of the fire problem for policymakers who must decide on allocations of resources between fire and other pressing problems, and within the fire service to improve codes and code enforcement, training, public fire education, building technology, and other related areas.

The Fire Administration, which has no regulatory authority, sends an experienced fire investigator into a community after a major incident only after having conferred with the local fire authorities to insure that the assistance and presence of the USFA would be supportive and would in no way interfere with any review of the incident they are themselves conducting. The intent is not to arrive during the event or even immediately after, but rather after the dust settles, so that a complete and objective review of all the important aspects of the incident can be made. Local authorities review the USFA's report while it is in draft. The USFA investigator or team is available to local authorities should they wish to request technical assistance for their own investigation.

This report and its recommendations were developed by USFA staff and by TriData Corporation, its staff and consultants, who are under contract to assist the USFA in carrying out the Fire Reports Program.

USFA wishes to acknowledge the support and cooperation of Donald O. Manning, Chief Engineer and General Manager; Donald F. Anthony, Deputy Chief; and Thomas E. McMaster, Battalion Chief, as well as many other members of the Los Angeles City Fire Department. Chief Anthony, the Incident Commander for this fire, provided invaluable information, supplied photographs and the fire department's own detailed report which served as a key reference particularly on the Los Angeles Incident Command System (ICS).

For additional copies of this report write to the U.S. Fire Administration, 16825 South Seton Avenue, Emmitsburg, Maryland 21727. The report is available on the Administration's Web site at http://www.usfa.dhs.gov/

# U.S. Fire Administration

## Mission Statement

*As an entity of the Department of Homeland Security, the mission of the USFA is to reduce life and economic losses due to fire and related emergencies, through leadership, advocacy, coordination, and support. We serve the Nation independently, in coordination with other Federal agencies, and in partnership with fire protection and emergency service communities. With a commitment to excellence, we provide public education, training, technology, and data initiatives.*

# TABLE OF CONTENTS

# FIRST INTERSTATE BANK BUILDING FIRE
## Los Angeles, California
## May 4, 1988

Local Contacts:    Donald F. Anthony
Deputy Chief
Thomas E. McMaster
Battalion Chief
Los Angeles City Fire Department
200 North Main Street
Los Angeles, California
(213) 485-6237

## OVERVIEW

On Wednesday, May 4, and continuing into May 5, 1988, the Los Angeles City Fire Department responded to and extinguished the most challenging and difficult highrise fire in the city's history. The fire destroyed four floors and damaged a fifth floor of the modern 62 story First Interstate Bank building in downtown Los Angeles. The fire claimed one life, injured approximately 35 occupants and 14 fire personnel, and resulted in a property loss of over 50 million dollars.

This was one of the most destructive highrise fires in recent United States history. The fire presented the greatest potential for the "Towering Inferno" scenario of any U. S. fire experience and was controlled only through the massive and dedicated manual fire suppression efforts of a large metropolitan fire department. It demonstrated the absolute need for automatic sprinklers to provide protection for tall buildings.

A total of 383 Los Angeles City Fire Department members from 64 companies, nearly one-half of the on-duty force of the entire city, were involved in fighting the fire, mounting an offensive attack via four stairways. This operation involved many unusual challenges, but is most notable for the sheer magnitude of the fire and the fact that the fire was successfully controlled by interior suppression efforts. To cover areas of the city protected by units called to the fire, 20 companies from Los Angeles County and four companies from surrounding jurisdictions were brought in under mutual aid agreements.

1

# SUMMARY OF KEY ISSUES

| Issues | Comments |
| --- | --- |
| Occupancy | 62-story highrise office building. |
| Construction | Steel frame; exterior of glass and aluminum. |
| | Unusually good application of fire resistive coating helped maintain structural integrity in fire. |
| Delayed Reporting | Building security and maintenance personnel delayed notifying fire department for 15 minutes after first evidence of fire. |
| | Smoke detectors on several floors had been activated and reset a number of times before reporting to fire department. |
| | A maintenance employee died while trying to investigate source of alarms prior to calling fire department. |
| Automatic Fire Sprinklers | Sprinkler system was installed in 90 percent of the building, including on fire floors; valves controlling the systems had been closed, awaiting installation of water flow alarms. |
| Interior Design and Contents | Large open area with readily combustible contents contributed to quick fire growth. |
| System Failures | Main fire pumps had been shut down, reducing available water pressure for initial attack. |
| | Radio communications were overtaxed and disrupted by building's steel frame. |
| | Fire and water damaged telephone circuits making them unusable. |
| | Sound-powered emergency phone system in building was ineffective. |
| Firefighter Safety | Fourteen firefighters sustained minor injuries in this operation, out of a total of nearly 400 involved. |
| | Use of protective hoods was effective in preventing neck and ear burns and allowing firefighters to penetrate fire floors. |
| | Attack crews used only 30-minute self-contained breathing apparatus (SCBA) to control their time on involved fire floors and avoid over fatigue. |
| | The Los Angeles City Fire Department attention to physical fitness is credited with minimizing effects of fatigue. |

## PREFACE

The Los Angeles City Fire Department provided information necessary to prepare this report through the generous cooperation of Chief Engineer and General Manager Donald O. Manning and Deputy Chief Donald F. Anthony.

The report is based heavily on information from a comprehensive report prepared by Battalion Chief Thomas E. McMaster, who was assigned to be the Los Angeles City Fire Department's historian for this fire.

The Los Angeles City Fire Department has the policy of assigning a chief officer to be the official historian for major fires. The purpose of this assignment is to document the facts surrounding significant fires for informational and training purposes. Through a careful analysis of the cause, progress, and extinguishment of major fires, the department identifies the need for improved standard operating procedures and legislation. The documentation effort is a comprehensive analysis of the fire with pertinent observations and recommendations which will be placed on the agenda of various fire department committees for further consideration.

## BUILDING DESCRIPTION

The First Interstate Bank building, the tallest in the city (and the State of California), is located at the intersection of Wilshire Boulevard and Hope Street in downtown Los Angeles. It was built in 1973, one year before a highrise sprinkler ordinance went into effect, and had sprinkler protection only in the basement, garage, and underground pedestrian tunnel. The 62-story tower measures 124 feet by 184 feet (22,816 square feet). It contains approximately 17,500 net square feet of office space per floor, built around a central core. It is occupied primarily by the headquarters of the bank corporation with several floors occupied by other tenants. Approximately 4,000 people work in the building.

The tower contains four main stairways (numbered 5, 5-A, 6, and 6-A in the 12th-floor plan in Appendix A). Stairs 6 and 6-A are enclosed within a common shaft, and stairway 5-A has a pressurized vestibule separating each floor with the stair shaft. Each stairway contains a combination standpipe with a pressure reducing valve at each landing. The building is topped with a helicopter landing pad.

The building has a structural steel frame, protected by a sprayed-on fire protective coating, with steel floor pans and lightweight concrete decking. The exterior curtain walls are glass and aluminum.

A complete automatic sprinkler system costing 3.5 million dollars was being installed in the building at the time of the fire. The installation was not required by codes at the time the owners decided to provide increased fire protection for the building. The project was approximately 90 percent complete, with work in progress at the time of the incident. The piping and sprinkler heads had been installed throughout the five fire floors and connected to the standpipe supply. However, a decision had been made to activate the system only on completion of the entire project, when connections would be made to the fire alarm systems, so the valves controlling the sprinklers on completed floors were closed.

## THE FIRE

The fire originated in an open-plan office area in the southeast quadrant of the 12th floor. (See Appendix A.) The area of origin contained modular office furniture with numerous personal computers and terminals used by securities trading personnel. The cause is thought to be electrical in origin, but the precise source of ignition was not determined. The fire extended to the entire open area and several office enclosures to fully involve the 12th floor, except for the passenger elevator lobby, which was protected by automatic closing fire doors.

The fire extended to floors above, primarily via the outer walls of the building; windows broke and flames penetrated behind the spandrel panels around the ends of the floor slabs. The curtain wall construction creates separations between the end of the floor slab and the exterior curtain wall. (For a discussion of this type of fire spread see National Fire Protection Association (NFPA) *Fire Journal*, May/June 1988, pages 75-84.)

There was heavy exposure of flames to the windows on successive floors as the fire extended upward from the 12th to 16th floors. The flames were estimated to be lapping 30 feet up the face of the building. The curtain walls, including windows, spandrel panels, and mullions, were almost completely destroyed by the fire. There were no "eyebrows" to stop the exterior vertical spread, and fireground commanders were concerned about the possibility of the fire "lapping" higher to involve additional floors.

Minor fire extension also occurred via poke-through penetrations for electricity and communications, via heating, ventilation, and air conditioning (HVAC) shafts, and via heat conduction through the floor slabs. A minor fire occurred in a storeroom on the 27th floor, ignited by fire products escaping from an HVAC shaft that originated on the 12th floor. This fire self-extinguished due to oxygen deficiency, but could have greatly complicated the situation if it had continued to burn. The secondary extensions were minor compared to the perimeter fire spread at the curtain walls.

The fire extended at a rate estimated at 45 minutes per floor and burned intensely for approximately 90 minutes on each level. This resulted in two floors being heavily involved at any point during the fire. The upward extension was stopped at the 16th floor level, after completely destroying four and one-half floors of the building.

## INITIAL STAGES

At 2222 on the night of the fire, the building's two fire pumps were shut down by the sprinkler contractor, and the combination standpipe system was drained down to the 58th floor level to facilitate connecting the new sprinkler system to the standpipe at that level. Three minutes later, at 2225, employees of the sprinkler system contractor heard glass falling and saw light smoke at the ceiling level on the 5th floor. A manual alarm was pulled but sounded for only a few seconds. It is believed that the alarm was silenced by security personnel on the ground floor.

At 2230, a smoke detector on the 12th floor was activated and was reset by security personnel. At 2232, three additional smoke detectors on the 12th floor were activated and were again reset by security personnel. At 2234, four smoke detectors on the 12th floor were activated and reset.

At 2236, multiple smoke detector alarms from the 12th to the 30th floors activated. A maintenance employee took a service elevator to the 12th floor to investigate the source of the alarms. The employee died when the elevator door opened onto a burning lobby on the 12th floor.

## FIRE DEPARTMENT OPERATIONS

At 2237, the fire department received three separate 9-1-1 calls from people outside of the First Interstate building reporting a fire on the upper floors. At 2238, a Category "B" assignment was dispatched consisting of Task Forces 9 and 10, Engine 3, Squad 4, and Battalion 1 – a total of 30 personnel. (A task force in Los Angeles consists of 10 personnel operating two pumpers and one ladder truck.)

The first report of the fire from inside the building was received at 2241, as the first due companies were arriving at the scene. While en route, Battalion 1 had observed and reported a large "loom-up" in the general area of the building. As he arrived on the scene, the battalion chief observed the entire east side and three-fourths of the south side of the 12th floor fully involved with fire. Battalion Chief Don Cate immediately called for five additional task forces, five engine companies, and five battalion chiefs. This was followed quickly by a request for an additional five task forces, five engine companies, and five battalion chiefs, providing a total response of over 200 personnel within five minutes of the first alarm. Two fire department helicopters were also dispatched.

The Highrise Incident Command System was initiated with companies assigned to fire attack and to logistics and support functions from the outset. Appendix B shows the system and indicates the changes in command of various functions during the course of the incident.

Appendix C shows the overall site and the location of the Command Post.

In accordance with Los Angeles City Fire Department policy, elevators were not used and all personnel climbed the stairs to the fire area. The first companies to reach the fire floor found smoke entering all four stairshafts from around the exit doors. Handlines were connected to the standpipe risers and the initial attack began at approximately 2310. Due to the magnitude of the fire on the 12th floor, attack was initiated from all four stairways. The crews had great difficulty advancing lines through the doors and onto the floor. As the doors were opened, heat and smoke pushed into the stairways and rose rapidly to the upper levels of the building.

The first six arriving companies were sent immediately to attack the fire. The initial attack used primarily 2-inch attack lines. The attack was hampered by low water pressure for the first few minutes, until the building fire pumps were started. The standpipes were also supplied by three fire department pumpers via the exterior hose connections.

As the attack was put into operation, a staging area was established on the 10th floor and a lobby control was established at ground level. The "base" for the operation was located a block south of the building (see Appendix C), following the Highrise Incident Command Plan.

The command post was established by the first arriving battalion chief one-half block south of the fire, and the incident commander operated from this location for the duration of the incident. The operations chief went to the 10th floor staging area to direct interior suppression efforts where he would have direct contact with officers assigned to each floor.

The key positions in the command organization were initially assigned to captains or battalion chiefs who were later relieved by higher ranking command officers. In most cases the relieved battalion chief stayed at the same location to work as an assistant to the higher ranking officer.

Deputy Chief Donald F. Anthony, second-in-command of the Los Angeles City Fire Department, became the Incident Commander upon his arrival at the scene. Chief Engineer and General Manager Donald O. Manning was also present at the command post and was involved in strategic planning for the incident.

## EXPANDING OPERATIONS

It soon became evident from the exterior and the interior that the fire was spreading upward. Companies successively launched attacks from all four stairways onto the 13th, 14th, 15th, and 16th floors, often encountering heavy fire from the point of entry and having to fight their way onto the floors with handlines. At times active suppression efforts were underway simultaneously on four levels as crews attempted to push the fire back from the central core to the perimeter of each floor. As more doors were opened, conditions in the stairways deteriorated with heat and smoke going up and water cascading down. (Appendix D shows a vertical cross-section of the building and the fire floors.)

The operations chief communicated with the command officers assigned to the floors above, directing tactical activities and making assignments of fresh or recycled companies to specific floors and stairways. Several companies handled three or four different firefighting assignments as conditions changed during the incident, with only short breaks to change air cylinders at the 10th floor staging area.

The operations chief relied primarily on runners to communicate with the floors above because radio communications were overtaxed and disrupted by the building's steel frame. In order to communicate with the Incident Commander, who was located on the street level, a window was broken out and a battalion chief stood at the opening with a portable radio to provide line-of-sight communications between the 10th floor and the command post. Attempts to use regular telephone service were unsuccessful due to fire and water damage to telephone circuits. An installed sound-powered emergency phone system, linking all floors with the lobby, also proved to be inadequate.

The strategy employed to stop the upward progress of the fire was to use aggressive tactics on the 14th and 15th floors to reduce the fire's intensity and the resulting exposure to floors above, while setting-up with hoselines and waiting for the fire to attack the 16th floor. This strategy proved to be successful but required extreme efforts by crews operating handlines on heavily involved floors, with as many as four floors burning below them. Approximately 20 handlines were used by 32 attack companies on the five involved floors.

## LOGISTICS

The logistical considerations involved in this operation were massive. The 10th floor was used as the staging area for personnel and equipment. Crews would return to the staging area to rest and change air cylinders and then to await reassignment to a fire floor. Companies went into action with full air cylinders and returned to the staging area when they were out of air. The companies operated for approximately 20 minutes in each cycle and had approximately 20 minutes to rest and change air cylinders.

Without elevators, every piece of equipment had to be carried up the stairs, including approximately 600 air cylinders. Every firefighter entering the building carried hose, nozzles, and other tools up to the 10th floor. A stairwell support operation, with nine companies assigned, spent over two hours moving equipment from the street level, through an underground tunnel from a parking garage across the street, up to the lobby, and then up the stairs to the staging area.

Crews working below the fire worked under deteriorating, adverse conditions. Smoke began to fill the 10th and lower floors, and windows had to be broken for ventilation. Water cascaded down the stairways and through the ceilings; and electrical power, including emergency lighting in the stairs, was lost. The operation continued for so long that even handlights failed due to battery consumption.

## SEARCH AND RESCUE

Approximately 50 occupants were inside the building and above the 12th floor when the fire broke out. These occupants including cleaning and maintenance workers, the sprinkler fitters, and a few tenants who were working late in offices. The occupants became aware of the fire as smoke entered the areas where they were working.

Five of the occupants from upper floors went to the rooftop and were rescued by helicopters. Others attempted to exit via elevators, some successfully and some unsuccessfully. At least one group found themselves on the 12th floor and had to crawl to an exit stairway in dense smoke and heat. Most of the occupants successfully exited via the stairs and encountered firefighters coming up the stairs as they descended.

The fire department was able to account for all except three known occupants of the building by comparing names with the sign-in sheets maintained by security personnel. Two of these occupants were on the 37th floor and one was on the 50th floor. Helicopter crews were able to locate all three at windows.

Due to the heavy smoke and heat conditions in all four stairways, it was impossible to send search and rescue crews to the upper floors until the fire was knocked-down at 0219. The Airborne Engine Companies were unable to penetrate from the rooftop until this time and were successful in rescuing the 50th floor occupant at approximately 0230.

Crews working from below found the two occupants of the 37th floor at approximately the same time. One of these occupants was unconscious and had to be carried down the stairs to the ground level.

## MEDICAL GROUP

A medical group was established by the Los Angeles City Fire Department one block east of the fire. Ten fire department paramedic ambulances, 17 private ambulances, and two hospital disaster teams were dispatched to this location. Approximately 37 building occupants and 14 fire department members were treated for injuries, primarily smoke inhalation and exhaustion. The only serious injuries were occupants of the building suffering from severe smoke inhalation, who were admitted to hospitals for observation.

## HELICOPTERS

The Los Angeles City Fire Department dispatched a total of four helicopters to the scene. The first two were automatically dispatched when the responding battalion chief reported a working highrise fire. One of these picked up a pre-designated "Airborne Engine Company" at a fire station while en route to the fire. A second "Airborne Engine Company" was requested early in the incident, and both were on the scene within 30 minutes.

An Air Operations Group, commanded by a battalion chief, was set up approximately eight blocks from the fire in an open area. The primary mission of this group was to land crews on the rooftop helipad to attempt search and rescue from the top down. Due to the heat and smoke that were encountered in the stairways, these crews were unable to descend more than two floors until the fire was controlled below. With so many doors open to provide access for firefighting, the stairways functioned as chimneys.

The helicopters made several trips to the roof to deliver and retrieve personnel. In the later stages of the fire, two representatives from the elevator service company were also taken to the roof. The helicopters also circled the building, reporting on exterior conditions and looking for occupants who might be visible on upper floors.

Prior to the arrival of the fire department helicopters, five persons were removed from the rooftop helipad by police department helicopters. The evacuees included two employees of the sprinkler contractor, one of whom responded back to the scene to turn on the fire pumps.

## FIREFIGHTER SAFETY

The Los Angeles City Fire Department places major emphasis on the safety of its personnel. In this incident the dangers to personnel were obvious and safety was a primary concern throughout the incident. A battalion chief was assigned as the safety officer and was directly involved in maintaining crew accountability.

The situation inside the building presented the risk of crews overextending their penetration and finding themselves too far above the staging area or too far into a burning floor to reach a safe area before their air supply was exhausted. Crews had to return to the 10th floor staging area to rest and change air cylinders before being reassigned to an active fire floor. A decision was made to use only 30-minute air cylinders, although 60-minute cylinders were available, to keep crews from overextending their penetration. Only the crews that went to the roof by helicopter used the 60-minute cylinders.

Fatigue was also a major concern, as several of the crews went through four cycles of fire attack, rest, replace cylinders, and return to action, after climbing 10 floors carrying all of their equipment.

As conditions deteriorated in the staging area, it was necessary to break windows for ventilation, and difficulty was encountered in providing even drinking water for the resting crews.

All of the air cylinders used at the scene were carried up the stairs, while empty cylinders were stockpiled on the 10th floor. Over 600 full cylinders were delivered to the scene and carried up to the staging area. The logistical problems of two-way movement precluded refilling cylinders at street level.

Some smoke inhalation and exhaustion injuries were reported, but none were serious. These occurred primarily with crews trying to conserve air, not using their SCBA until they reached their assigned floors or running out of air before reaching the staging area. The combination of stress, smoke, and fatigue was extremely challenging, and the fact that no serious injuries occurred is a credit to the training and physical fitness of Los Angeles firefighters.

The heat conditions encountered by attack crews were severe, and the use of protective hoods was credited with preventing ear and neck burns as the crews advanced lines into involved floors. No burn injuries were reported.

## FALLING GLASS

Falling glass and other debris created a major problem during this incident. Virtually all of the exterior curtain wall, from the 12th through 16th floors, was destroyed and fell to the ground. The falling glass and debris caused significant damage to pumpers hooked up to the fire department connections. The hoselines were cut several times and had to be replaced, under the constant danger of additional falling materials. The entire perimeter of the building, for over 100 feet out from the walls, was littered with this debris.

Fortunately, a tunnel between the building and the parking garage across the street provided a safe path into the building at the basement level for both personnel and equipment. Without this tunnel it would have been very difficult to maintain the necessary logistical supply system and to avoid injuries to personnel from the falling debris.

The windows were coated with a plastic reflective material which caused them to hold together as they fell. The glass fell in very large sections, some of which were flaming due to ignition of the plastic coating.

The Incident Commander gave blanket approval for crews to break windows to provide ventilation, since glass and debris were already falling on all four sides of the building. Firefighters reported difficulty in breaking the thick windows, and the coating on the windows may also have been a factor in this regard. A pick-head axe was found to be the most effective tool for the job. (A police sharpshooter offered to shoot windows out from a helicopter, but the offer was declined.)

## SPRINKLERS AND STANDPIPE SYSTEM

The building was served by a single zone combination standpipe system with four risers, one in each stairway. The standpipe risers provided a 2-1/2-inch outlet in each stairway at each floor and also supplied 1-1/2-inch hose cabinets in the occupied areas of each floor. At the time the fire broke out the building's fire pumps had been shut down, and the risers had been drained down below the 58th floor to allow a sprinkler line to be connected to a standpipe riser. This job took only a few minutes and the jockey pump was in operation to refill the system when the sprinkler crews noted smoke rapidly filling the stairway. Being only four floors from the roof, the workers went up to the helipad to await rescue.

The sprinkler system was virtually complete on the floors that burned, but the valves were closed between the standpipe riser and the sprinkler system on each floor. During the fire a battalion chief was assigned to confer with the sprinkler installation supervisor to explore the possibility of opening these valves to control the fire. It was determined that the fire on the involved floors would probably overwhelm the sprinklers and deprive the handlines of needed water. Eventually, the systems on floors 17, 18, and 19 above the fire were activated, in case the fire extended past the 16th floor.

The four standpipes are supplied by two stationary fire pumps, one diesel and one electric, each rated at 750 gpm at 600 psi pressure. The standpipe system operates with a single vertical zone, depending on the pressure reducing valves at each outlet to control the pressure.

With the main fire pumps shut down, the only water pressure available for the first hoselines was the static head in the risers, and crews reported poor water pressure for the first few minutes. This condition was rapidly corrected when the combination of both building pumps and three fire department pumpers were placed in operation.

The building pumps were started manually by the sprinkler installation supervisor who had been rescued from the rooftop by a police helicopter, taken to a police facility, and transported back to the scene in a police car. He arrived at the fire department command post and informed the Incident Commander of the situation. An engine company was assigned to drive him into the basement loading dock area in a car, to avoid the falling glass, and to assist him in starting the pumps.

The building's two 750 gpm fire pumps drew water from an 85,000 gallon reservoir in the sub-basement. The resupply from the public water supply system was unable to keep pace with the outflow, estimated at over 2,000 gallons per minute, and there were fears that the tank would be emptied. The tank was down to less than one-third of its capacity when the fire was controlled. If the tank had emptied, only fire department pumpers would have been left to supply the standpipes.

The single zone riser system was designed to operate at 585 pounds per square inch (at basement pump discharge) and relied upon the pressure reducing valves to limit the discharge pressure at each outlet on each landing. Problems were encountered with several of these valves allowing excess pressure to be discharged, including one that provided over 400 pounds per square inch. The overpressure caused several hose ruptures and made handlines difficult to control. The heat of the fire caused several aluminum alloy valves in the occupant hose cabinets to fail, creating high pressure water leaks. These leaks took water from the supply that was available for handlines and caused additional water damage on floors below the fire.

It was estimated that a total flow of 4,000 gallons per minute was delivered by the standpipe risers. The total effective fire flow, provided by hoselines attacking the fire, was approximately 2,400 gpm. The attack lines included 1 3/4, 2, and 2 1/2 inch handlines. No exterior streams or master stream appliances were used.

## ELEVATORS

No elevators were used by the fire department during this incident. There is a standard policy in the Los Angeles City Fire Department **not** to use elevators that have a shaft opening within five floors of the fire floor. The use of any elevators was ruled out based on the amount of fire that was visible from the street.

The building contained 31 passenger elevators in four banks above the main lobby, two banks for sublevels, and two service (freight) elevators. Both service elevators served all levels and were provided with a lobby at each floor. All of the elevators were designed to be recalled to the ground floor lobby on smoke detector activation. Some elevator cars were returned to the ground floor, but those that had been in use by cleaning and maintenance crews were on "Independent Service" and could not be recalled. Other cars were not accounted for because their doors did not open when they returned to the ground floor. One elevator stopped on the 22nd floor and one stopped on the 33rd floor. This caused concern that individuals could be trapped in the 10 "missing" elevators and attempts were made to locate these cars. Their locations were eventually determined by the elevator maintenance personnel who were taken to the rooftop machine room by helicopter.

The single fatality was found in a service elevator. The building employee, who went to the 12th floor to investigate the alarm, was trapped and died when the elevator opened into a burning lobby. The lobby fire separation was compromised by a cleaner's cart blocking the door open. The victim was able to call for help on his portable radio, but other employees had no means to rescue him. This elevator was put on independent service through the use of a key by the building employee.

Lobby isolation doors had been installed on the passenger elevator lobbies of several floors, including the 12th and were successful in keeping fire and smoke out and providing a potential area of refuge. The isolation doors were being installed as the building was renovated. The fire reached the elevator shafts on floors where the doors had not been installed, and they became additional vertical conduits for heat and smoke. The machine room on the 22nd floor, which served the "low rise" elevators, received extensive heat and smoke damage. Smoke traveled throughout the building above the 12th floor via the elevator shafts.

## SALVAGE AND PROPERTY LOSS

The floors below the fire received massive water damage, and those above were heavily damaged by heat and smoke. During the fire, no efforts were directed toward property conservation as all available firefighters were committed to stopping the progress of the fire.

After the fire, the building remained closed for several months while the structure was inspected, and a large force of clean-up contractors worked through the building. The efforts to save property were conducted on a very large scale as virtually every part of the building was damaged by flames, heat, smoke, or water. (As part of the cleanup, 250,000 cloth diapers were used.) The property loss has been estimated at over 200 million dollars, without taking into account the business interruption loss.

In spite of the total burnout of four and a half floors, there was no damage to the main structural members and only minor damage to one secondary beam and a small number of floor pans. Although there was concern for structural integrity during the incident, post fire analysis indicates that there was no danger of major or minor structural collapse. It was noted that quality control in the application of the sprayed-on fire protection was unusually good.

## LESSONS LEARNED

1.  **Sprinkler system: use the protection as soon as possible.**

    The value of automatic sprinklers in quickly controlling fire and preventing fires of this magnitude must be emphasized. If the sprinkler system had been activated as floors were completed, the fire probably would have been controlled in minutes with minor damage. As buildings are constructed, renovated, or demolished, sprinklers should be kept operational on all the floors possible. Many fires occur during these stages of the life cycles of buildings, and they often are severe.

    The city of Los Angeles contains over 750 highrise buildings, approximately 450 of which are not protected by automatic sprinklers. This fire provided the lesson that was necessary to have a retroactive sprinkler installation requirement adopted by the City Council.

2.  **Unsprinklered highrise fires create massive staffing requirements.**

    The fire took advantage of a large open area, with readily combustible contents, to quickly reach major proportions. This, combined with an available path for vertical spread, created a situation that taxed a large, well equipped, and experienced fire department to its maximum. Many potentially serious problems arose, such as failing standpipe valves and delayed activation of building fire pumps. A fire department without the resources, capabilities, and experience of the Los Angeles City Fire Department would have great difficulty controlling upward extension, if faced with the same circumstances.

3.  **High danger to firefighters was mitigated by physical fitness, good personal safety equipment, and safety training.**

    The fact that almost 400 fire department members operated on this fire, with only 14 minor injuries, is a credit to the training and physical fitness of Los Angeles firefighters and the safety procedures that were employed. The use of protective hoods was found to be very effective in preventing burns and allowing firefighters to penetrate into the involved fire floors.

4.  **ICS is critical for a large, complex fire.**

The Highrise Incident Command System was very effective in managing the incident. Despite the massive numbers of companies and firefighters on the scene, the fire department maintained good organization at the scene and effectively--and safely-- managed their resources. The Los Angeles City Fire Department is to be commended for its extraordinarily low injury rate at this fire.

5.  **Communications within and from a steel frame building still can be a problem.**

The difficulties that were experienced with radio communications will require additional attention. An operation of this magnitude involves a high demand for communications capacity. In addition, the sound powered telephone system was found to be inadequate (and completely compromised when the system wires melted). The Los Angeles City Fire Department is in the process of installing an 18 channel 800 MHz radio system to address these problems.

6.  **Radio communications can easily be overloaded without strict radio discipline and an adequate number of channels.**

One of the major problems was the over usage of fireground radio channels. Also, communications from the air operations and medical groups interfered with interior tactical communications.

7.  **Building personnel must be trained to take appropriate actions when alarms are activated.**

The actions of building security and maintenance personnel in the first minutes of this incident are a cause for concern. The alarm was delayed in reaching the fire department, occupants of the building were not notified of the fire, and a life was lost while building personnel attempted to verify the source of the alarms.

8.  **Fire-resistive structures can maintain structural integrity if built well.**

The structural integrity of the building was a concern during and after the fire. Analysis revealed that no significant damage occurred to major structural elements. Part of this credit must go to the unusually good application of fire resisting materials on support members. The effects of this magnitude of fire on a less protected structure must be considered in plans review, inspections during construction, and developing codes.

9.  **Protected elevators are needed for fire service use.**

The lack of elevators for delivering firefighter personnel and equipment was a problem in this fire, although it occurred at a relatively low level in the building. If firefighters had to climb 50 stories instead of 15, the problems would have been compounded. This points to the need for carefully planning higher level operations. Different elevator banks may allow limited use of elevators that do not open on any involved floors.

10. **Smoke in stairways is still a problem.**

The concept of maintaining at least one stairway free of smoke, to be used for evacuation, proved ineffective in this incident. This concept may be valid for a less severe fire, but when the fire reaches this magnitude all vertical shafts become potential chimneys. The ventilated vestibule design failed to keep heat and smoke out of the pressurized smoke tower.

11. **Fire departments should develop contingency plans that contemplate the failure of systems to perform as designed, especially for major buildings.**

Fire departments must contemplate operating in buildings where fixed fire protection and other systems fail to operate as planned. If the individual with specific knowledge of the building fire pumps had not arrived at the command post, the pumps might have remained inoperative. The fire also disrupted HVAC systems, communications, and electrical power supplies beyond previous experience with highrise fires.

12. **Vertical and horizontal fire spread can still be rapid in modern buildings without sprinklers and without adequate compartmentation.**

Vertical fire spread and fire development in open floor areas were major factors in this incident. The floor of origin might not have become involved as quickly if it had been divided into smaller offices, providing for more rapid control of the fire. Exterior features of building design can be provided to reduce the risk of vertical flame impingement. Automatic sprinklers are usually effective in dealing with both of these concerns.

13. **Old Lesson: Fire protection systems need to be tested regularly.**

All components of fixed fire protection systems, including items such as pressure reducing valves, must be regularly inspected and tested. The problems encountered with the standpipe pressure reducing valves in this building could have had a crippling effect on fire suppression efforts.

14. **Falling glass is a special hazard in highrise fires.**

This has been a common problem at major highrise fires such as the Prudential fire in Boston. Large sheets of glass can act as guillotines. The existence of a tunnel for safe entry of personnel was fortuitous in this fire. Plans for new highrises should be reviewed for protected access by emergency personnel. Pre-fire plans for existing highrises should be reviewed as to how the local fire department would cope with this hazard.

15. **A major highrise fire requires a heavy commitment of personnel to logistics functions.**

The Los Angeles Department had thought in terms of a 3 to 1 ratio between support troops and firefighting troops. The ratio needed turned out to be considerably less than that at 1 to 1, but still high.

16. **"Fire-proof" vaults worked well to save valuable papers.**

An estimated 100 million dollars in stocks and bonds were successfully protected in a fire-proof vault exposed to the fire.

17. **Building security personnel must be trained to promptly report fires.**

The security personnel are believed to have silenced the alarm systems and wasted time in going to investigate the source of the smoke alarm. This not only resulted in a fatality but undoubtedly led to the fire being much larger by the time it was reported to the fire department. The chain of alarms being set off was still not recognized as possibly a rapidly spreading, large fire. This is not the first highrise were security personnel have exhibited similar behavior. Fire departments should stress the importance of prompt reporting and remind building owners of the risks that are involved in delayed reporting – including litigation. Fire departments should also consider requiring automatic alarms to transmit to Central Station Monitoring Systems.

## REFERENCES

The key reference on this fire is the detailed report of the Los Angeles City Fire Department. It provides an analysis of how the Los Angeles ICS worked and how it could be improved, especially the details of communications.  It also contains many additional "lessons learned."

The magazine article in <u>Fire Engineering</u> provides additional details on the construction of the building, its strengths and weaknesses.  Los Angeles Deputy Chief Tim DeLuca, Operations Chief at the fire, gave an excellent illustrated presentation on the fire at the NFPA 1988 annual conference in Nashville, Tennessee, which was used for some additional details and recommendations in this report.

Elmer Chapman, "High-Rise:  An Analysis," <u>Fire Engineering</u>, August 1988.

Harvey Eisner, "Towering Inferno," <u>Firehouse</u>, October 1988.

Donald Anthony, "Interstate Bank Building Fire," speech at NFPA Annual Conference, Nashville, TN, November 1988.

# APPENDICES

A. Twelfth-Floor Plan, showing area of origin and location of fatality.

B. Incident Command Organization.

C. Site Map, showing command post, operations bases, and medical group.

D. Vertical cross-section of building.

E. Copies of photographs provided by the Los Angeles City Fire Department.

# APPENDIX A

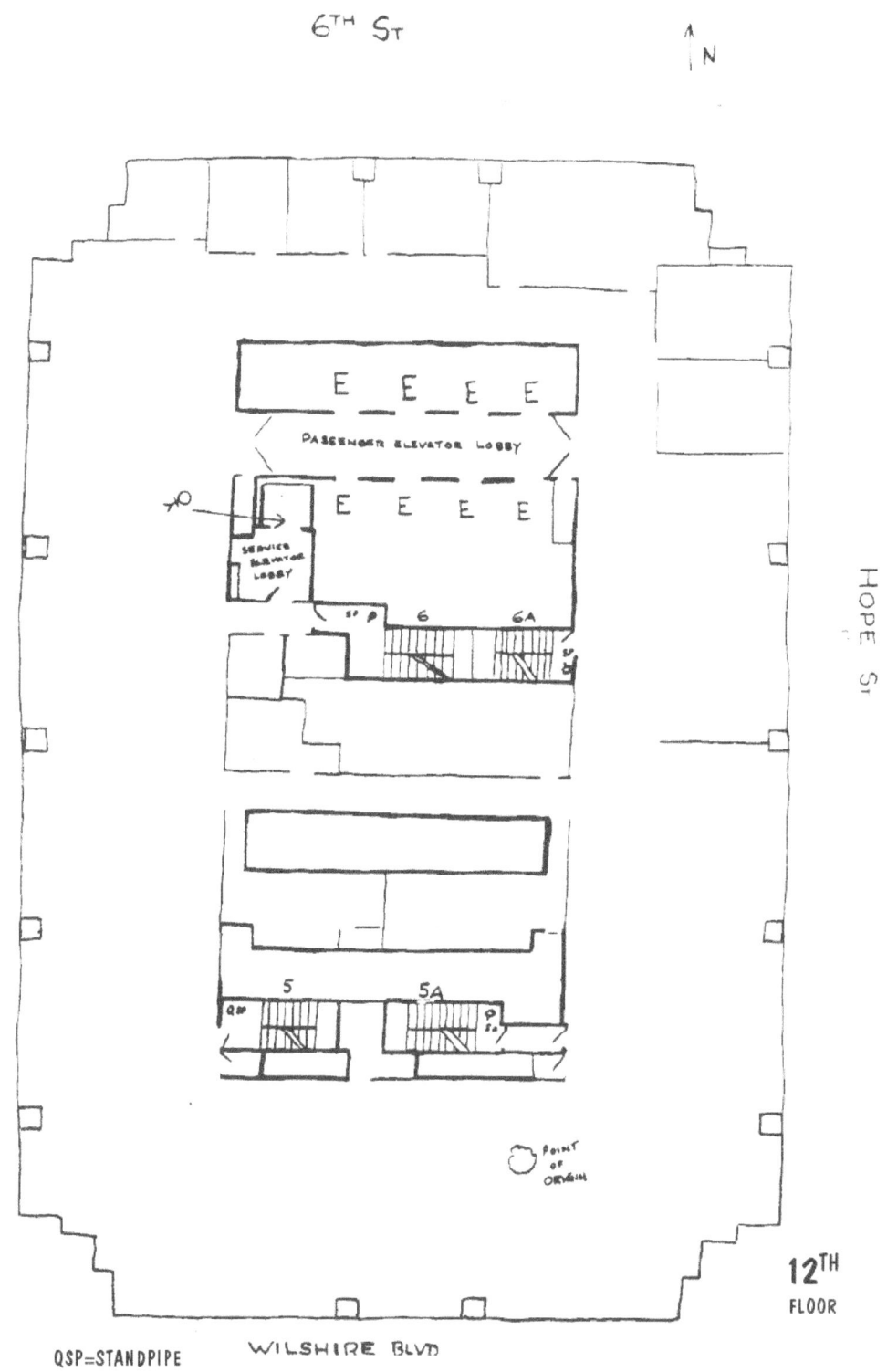

6TH ST

N

HOPE ST

PASSENGER ELEVATOR LOBBY

E  E  E  E

E  E  E  E

SERVICE ELEVATOR LOBBY

6  6A

5  5A

QSP

Point of Origin

12TH FLOOR

QSP=STANDPIPE

WILSHIRE BLVD

16

# APPENDIX B

## INCIDENT COMMAND ORGANIZATION
### FIRST INTERSTATE BANK BUILDING FIRE
MAY 4, 1988

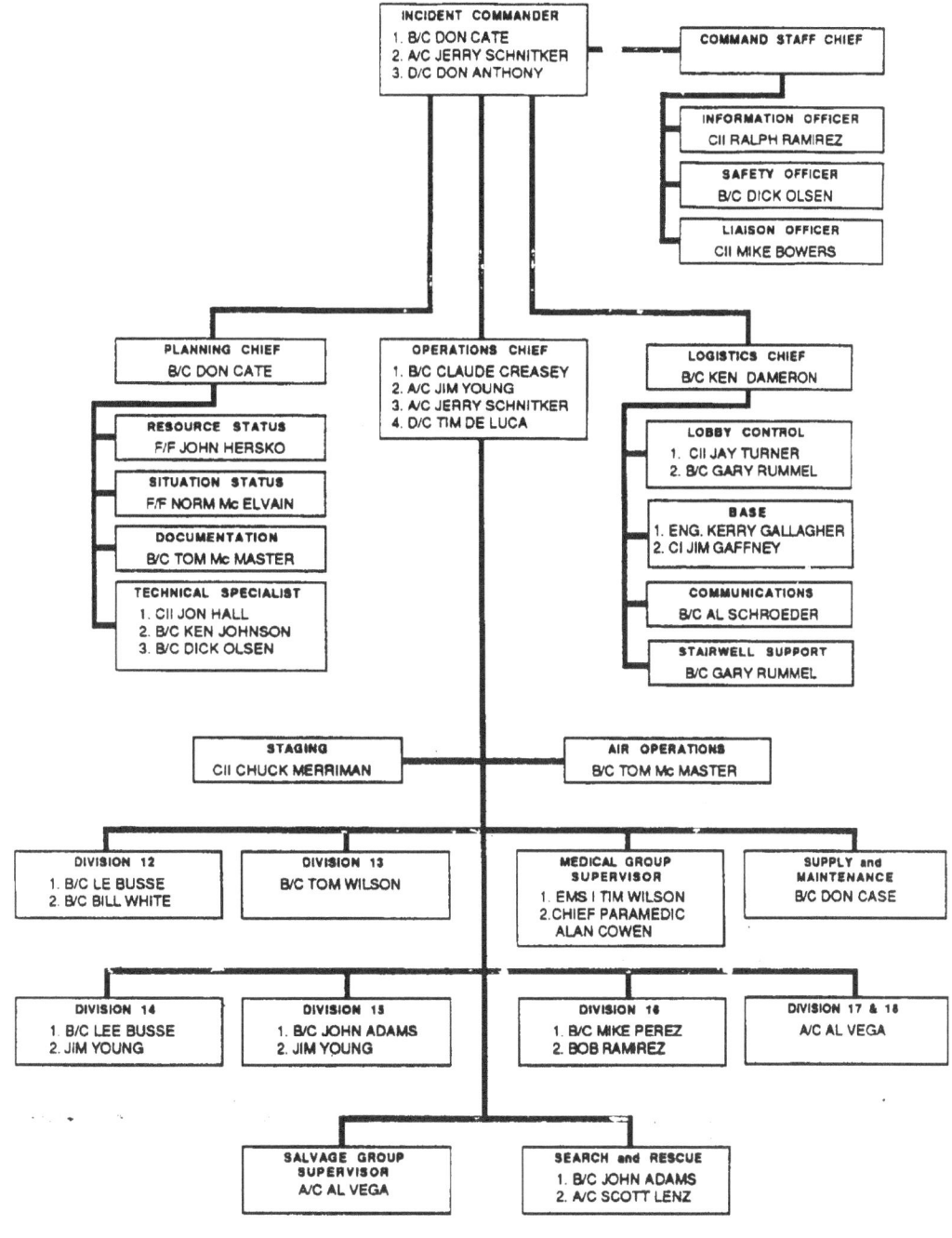

# APPENDIX C

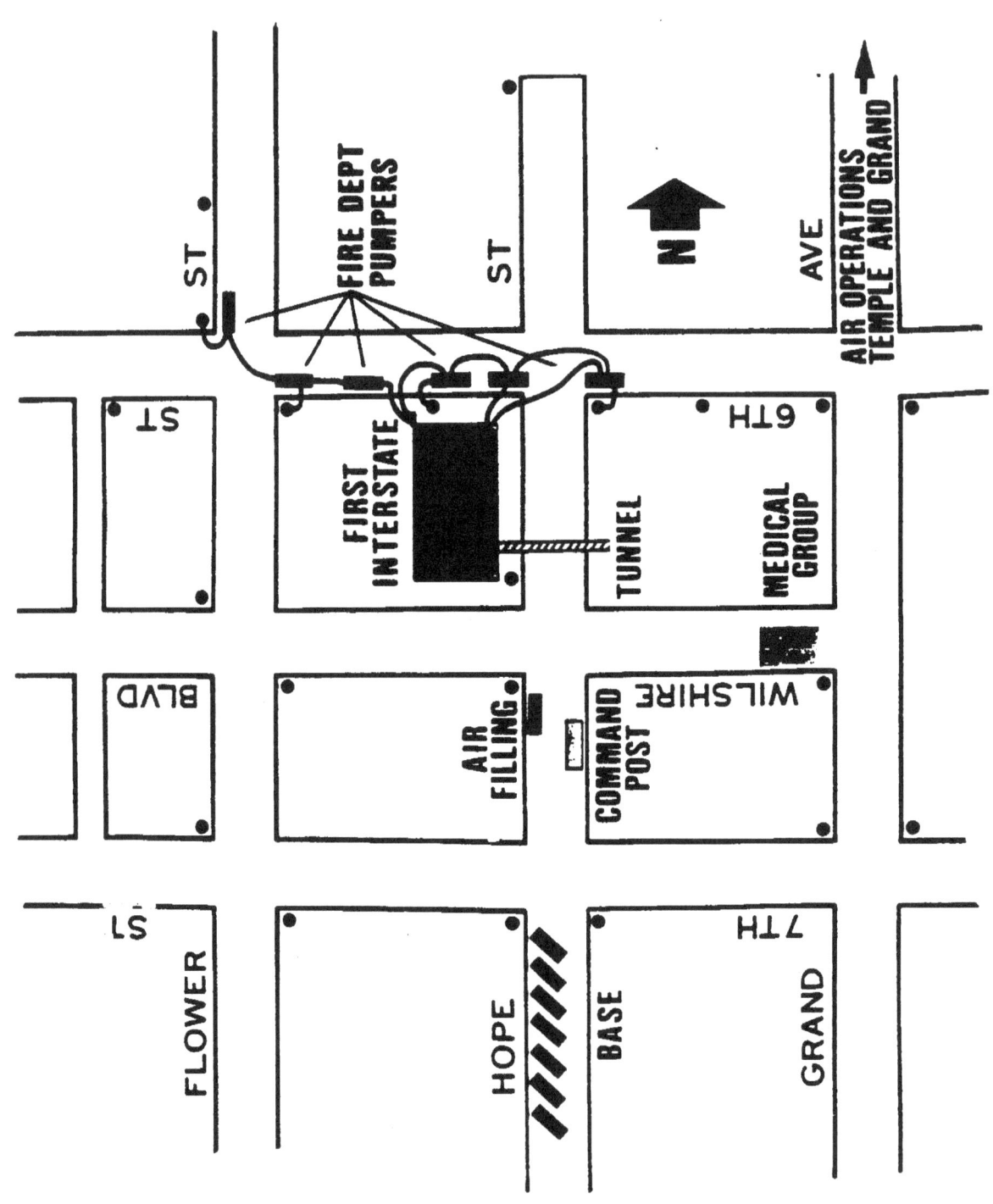

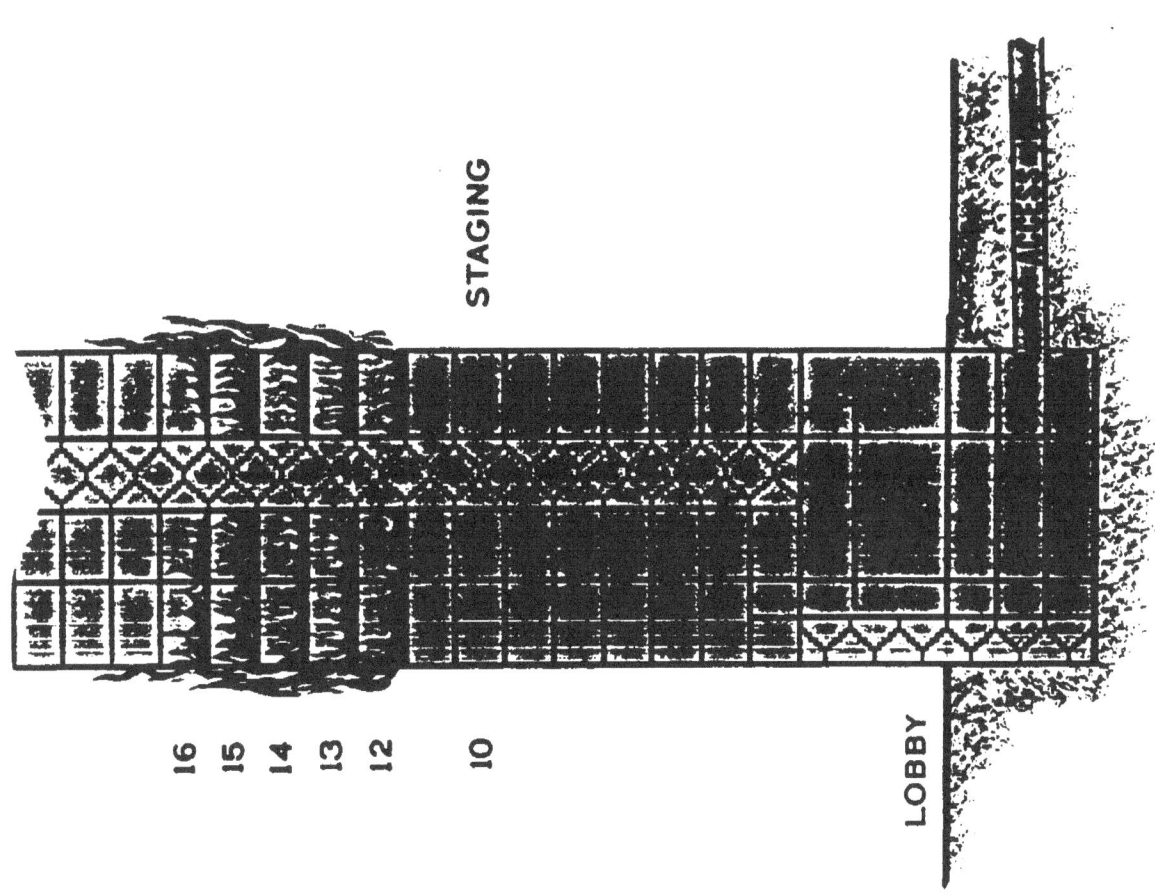

# APPENDIX E

Copies of photographs provided by the Los Angeles City Fire Department.

1.  Lobby control equipment pool.

2.  Empty air bottles in staging.

3.  Task Force 15 returning to staging – note hoods and water in stairwell.

4.  Tenth floor staging area.

5.  Helicopter searching 50th floor for trapped civilians.

6.  Wilshire Incident Command.

7.  Evidence of smoke in elevator shaft at 42nd floor.

8.  Stairwell 6 at 22nd floor showing smoke leakage.

9.  Twisted aluminum framework northeast corner of building.

10. Void between floor and exterior glass.

11. Plate glass windows on the 16th floor.

1. Lobby control equipment pool.

**2.  Empty air bottles in staging.**

3. Task Force 15 returning to staging – note hoods and water in stairwell.

4. Tenth floor staging area.

5. Helicopter searching 50th floor for trapped civilians.

**6. Wilshire Incident Command.**

7. Evidence of smoke in elevator shaft at 42nd floor.

**8. Stairwell 6 at 22nd floor showing smoke leakage.**

**9. Twisted aluminum framework northeast corner of building.**

**10. Void between floor and exterior glass.**

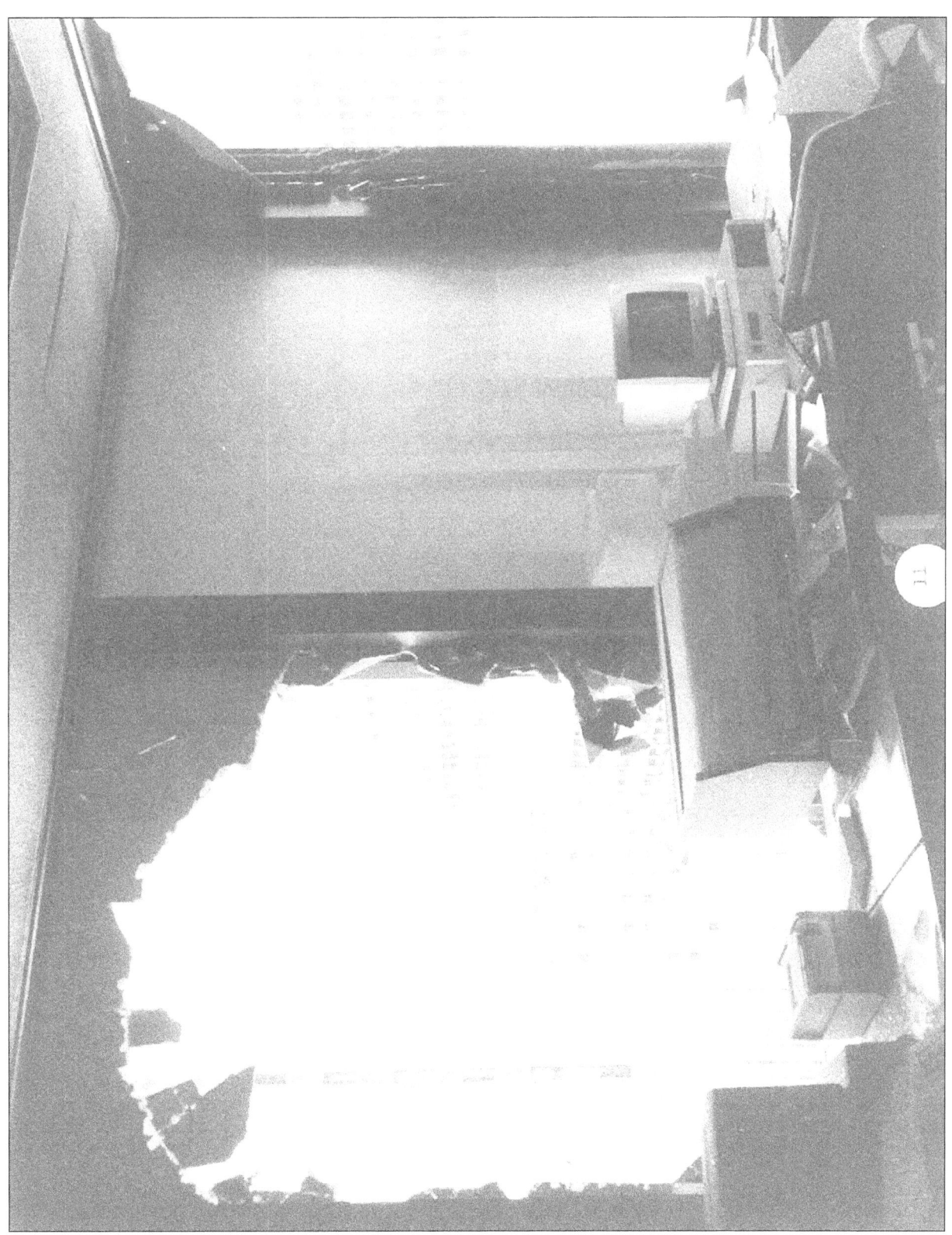

11. Plate glass windows on the 16th floor.